Grade 1

Acknowledgments
Product Development: Margaret Fetty
Editor: Michelle Howell
Cover Illustration: Tad Herr
Design and Production: Creative Pages, Inc.
Production Supervision: Sandy Batista

ISBN 1-59137-113-9
Options Publishing Inc.
P.O. Box 1749
Merrimack, NH 03054-1749
www.optionspublishing.com
Phone: 800-782-7300 Fax: 866-424-4056

15 14 13 12 11 10 9

Dear Parent,

Summer is a time for relaxing and having fun. It can also be a time for learning. *Summer Counts!* can help improve your child's understanding of important skills learned in the past school year while preparing him or her for the year ahead.

Summer Counts! provides grade-appropriate practice in subjects such as reading, language arts, vocabulary, and math. The ten theme-related chapters motivate your child to continue learning throughout the summer months.

When working through the book, encourage your child to share his or her learning with you. With *Summer Counts!* your child will discover that learning is a year-round process.

Apreciados padre,

El verano es una época para descansar y divertirse. También puede ser una época para aprender. *Summer Counts!* puede ayudar a que su hijo(a) mejore las destrezas importantes que aprendió el pasado año escolar al mismo tiempo que lo(a) prepara para el año que se aproxima.

Summer Counts! provee la práctica apropiada para cada grado en las asignaturas como la lectura, las artes del lenguaje y las matemáticas. Los diez capítulos temáticos incluyen actividades y rompecabezas que motivarán a su hijo(a) durante el verano.

Cuando trabaje con el libro, anime a su hijo(a) a que comparta lo que ha aprendido con Ud. Si Ud. desea puede desprender la página de las respuestas que aparece en la parte trasera del libro. Puede usar la misma para revisar el progreso de su hijo(a). ¡Con *Summer Counts!* su hijo(a) descubrirá que el aprendizaje puede ocurrir en cualquier momento—inclusive en el verano!

Table of Contents

CHAPTER 1

Hey, Diddle, Diddle

Hey, diddle, diddle!
The cat and the fiddle,
The cow jumped over the moon.
The little dog laughed
To see such sport,
And the dish ran away with the spoon.

Hey, Diddle, Diddle

Directions **Use what you have read. Answer the questions.**

1. What is a fiddle?

2. Who jumped over the moon?

3. Why do you think the dog was laughing?

4. Could this rhyme really happen? Why or why not?

5. How do you know that Hey, Diddle, Diddle! is a nursery rhyme? Circle one.

A. A nursery rhyme always has six lines.

B. A nursery rhyme is a short poem.

C. A nursery rhyme always makes sense.

Sing With Me

Directions **Write the missing letters.**

A B ______ D E F G
1

H ______ J K L M N O P
2

Q ______ S T U V
3

W X ______ Z
4

Now I know my ABCs, won't you sing along with me?

Make Words

Directions Draw lines to make the words from the box below. Write the words in the box shapes. The first one is done for you.

do	love	run	see	stop

1. l — op — love

2. r — ee

3. d — ove

4. s — o

5. st — un

Rhyming Words

Directions Read each word. Find the rhyming words in the box above. Write them on the lines.

6. shoe ____________ ____________

7. sun ____________ ____________

8. me ____________ ____________

9. top ____________ ____________

Counting Fun

REMEMBER

A **number** tells how many. A number can be written as a **word** too.

EXAMPLES

Number	Word	Number	Word
0	zero	5	five
1	one	6	six
2	two	7	seven
3	three	8	eight
4	four	9	nine

Write It Out

Directions **Write each number.**

1. 

____

2. 

____

3. 

____

Directions **Write each number as a word.**

4. 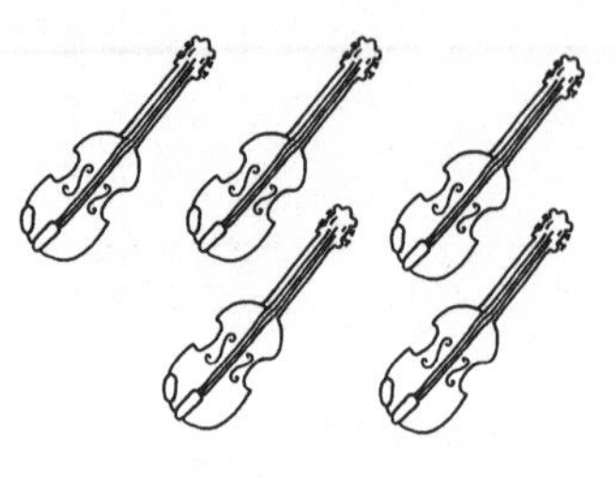

____

5. 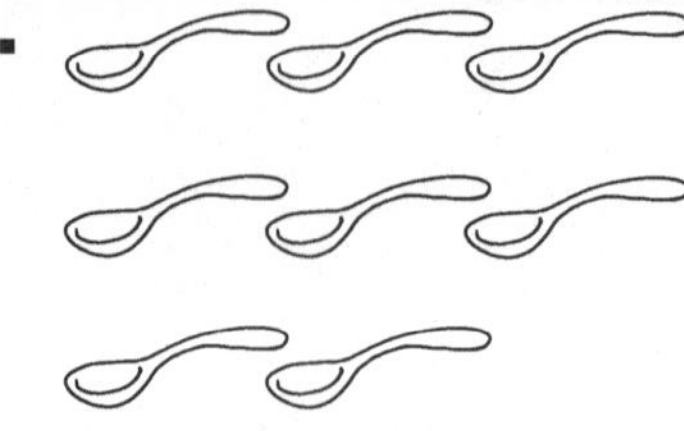

____

6. 

____

Counting Lines

REMEMBER

Tally marks are used for counting.
One line means there is one thing to count.

A group of 3 is shown as |||

A group of 5 is shown as ~~||||~~

How Many?

Directions Show each number using tally marks.

1.

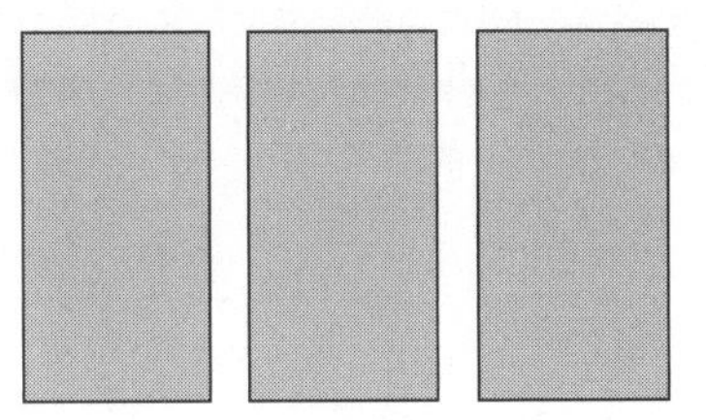

____

2.

____

3.

____

4.

____

5.

____

6.

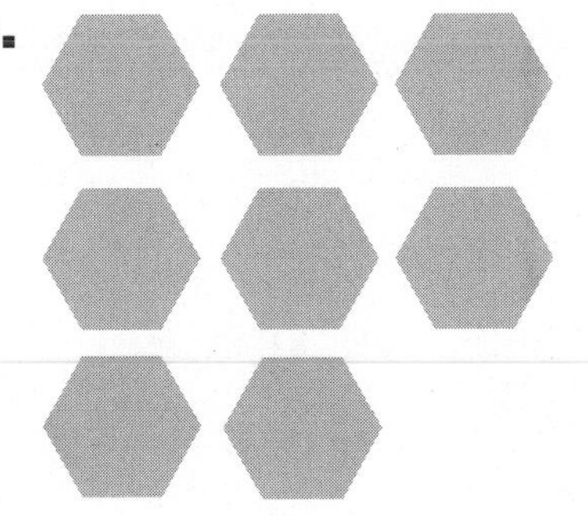

____

Hide and Seek Words

Directions **Circle the words listed in the box. The words can be across or down.**

hop	jump	skip	leap
run	play	walk	dance

t	s	k	i	p	m	i	l
r	u	n	h	l	r	s	e
r	d	u	d	a	n	c	e
v	l	s	e	y	c	o	j
s	e	h	o	p	a	g	u
i	a	e	c	a	u	w	m
c	p	e	w	a	l	k	p
d	z	p	t	s	h	u	l

Ready, Set, Jump!

Directions Look at the picture. What is happening? Write a story to go with the picture.

CHAPTER 2

COOK IT UP!

You can cook with someone.
Mix up a pancake.
Put it in a pan.
It will heat up.
Watch the pancake bubble.
Flip it one time.
Cook that side, too.
Soon it will be done.
Then you will have something
good to eat!

Cook It Up!

Directions **Use what you have read. Answer the questions.**

1. What is this story mostly about?

2. What would be another good title for this story?

3. What do you cook the pancake on?

4. What do you do after the pancake bubbles?

5. How do you think the pancake will taste?

In the Kitchen

Directions **Look at each picture. Circle the letter that stands for the first sound in each picture.**

1.	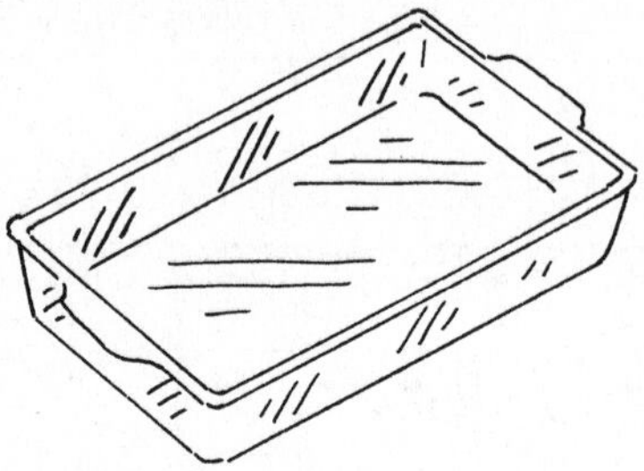 p	b	n
2.	 w	s	g
3.	 f	s	n
4.	 j	d	b

Kitchen Words

Directions Draw lines to make the words from the box below. Write the words in the box shapes.

box	cup	spoon	can	eat

1. ea **oon**

2. c **ox**

3. b **up**

4. sp **t**

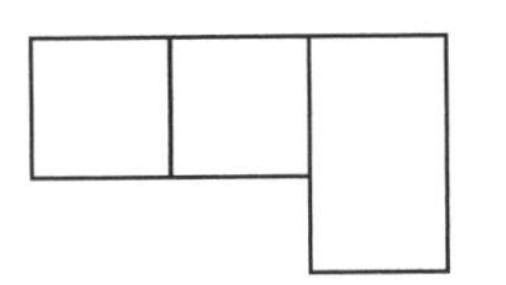

5. c **an**

Picture Clues

Directions Look at each picture. Find the words from the box above that name each picture. Write them on the lines.

6. ______________

7. ______________

8. ______________

9. ______________

Counting Tens

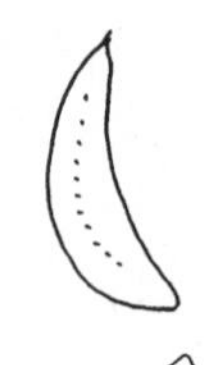

REMEMBER

Some numbers have both ones and tens.

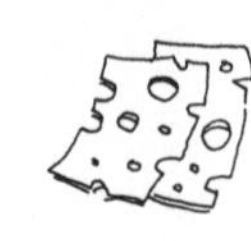

EXAMPLE

45 has 4 tens and 5 ones.

40 has (10) (10) (10) (10) (4 tens)

5 has (1) (1) (1) (1) (1) (5 ones)

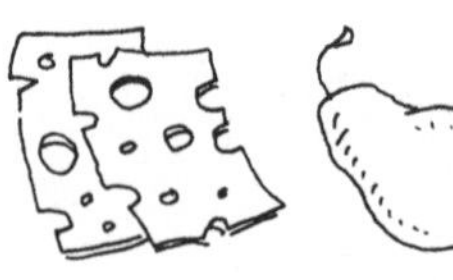

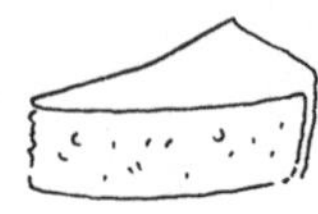

Counting Game

Directions Some children are playing a counting game. Count to see how many points each child has.

1. Sara

(10) (1)

____

2. Salim

(1) (10) (10) (1) (1)

____

3. Paul

(10) (1) (1) (10) (1) (10) (1)

____

4. Tia

(10) (10) (1) (1) (10) (10) (1) (1) (1)

____

Calendar Fun

Directions **Look at the calendar. Then answer the questions.**

JUNE

Sunday	Monday	Tuesday	Wednesday	Thursday	Friday	Saturday
1	2	3	4	5	6	7
8	9	10	11	12	13	14
15	16	17	18	19	20	21
22	23	24	25	26	27	28
29	30					

1. How many days are in June? ____________________

2. What day of the week is June 1st? ____________________

3. What day of the week is June 24th? ____________________

4. What month comes before June? ____________________

5. What day is after Friday? ____________________

6. How many Mondays are in June? ____________________

Food Scramble

Directions Read each meaning. Then unscramble the letters to find the words. Write the words on the lines.

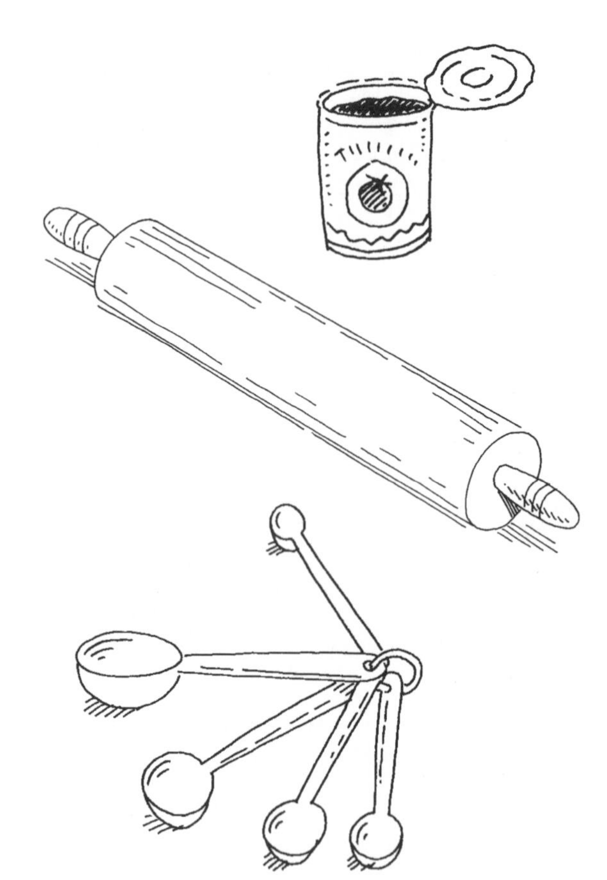

1. To chew food

tae ___ ___ ___
(1)

2. Something soup comes in

acn ___ ___ ___
(2)

3. Something to eat with

krof ___ ___ ___ ___
(3)

4. Something you eat off of

idsh ___ ___ ___ ___
(4)

Directions Answer the riddle. Match the letters above with the numbers. Write the letters on the lines.

What did the father corn say when his son wanted to talk?

I'm all ___ ___ ___ ___!
1 2 3 4

Your Turn to Cook

What food do you like to eat? Why do you like to eat it? Write about what it looks like and tastes like. Then draw a picture of your favorite food.

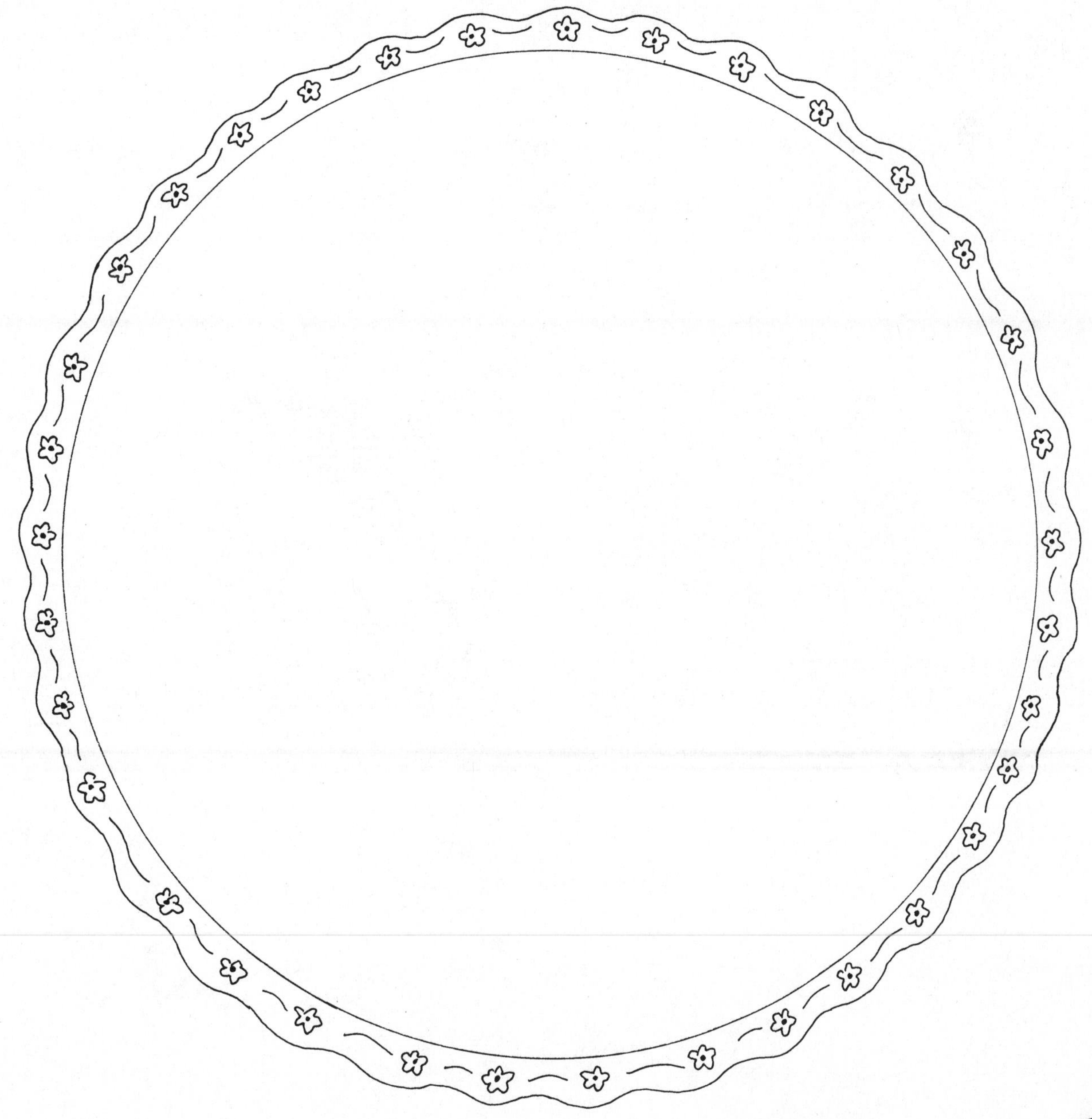

CHAPTER 3

Morning on the Farm

Cock-a-doodle-doo!
It is time to get up.
The farmer has work to do.
He will feed the animals.
Oink! He feeds the pigs.
Moo! He feeds the cows.
Cluck! He feeds the chickens.
All the animals have food.
Now it is time for the farmer to eat!

Morning on the Farm

Directions **Use what you have read. Answer the questions.**

1. Where does this story take place?

2. What animal wakes the farmer?

3. Who does the farmer feed first?

4. What does the farmer do after he feeds all the animals?

5. Could this story really happen? How do you know?

Animal Match-Up

Directions **Look at the letter on each grown-up animal. Circle the letter on the baby animal that matches it.**

1.	 T	l t k
2.	 M	m n w
3.	 O	c u o
4.	 P	g p p

Farm Animals

Directions Draw lines to make the words from the box below. Write the words in the box shapes.

cow	frog	sheep	pig	cat

Rhyming Words

Directions Read each word. Find the rhyming words in the box above. Write them on the lines.

6. now ______________ **7.** dog ______________

8. keep ______________ **9.** bat ______________

Counting Hundreds

REMEMBER

Some numbers have ones, tens, and hundreds.

EXAMPLE

325 has 3 hundreds, 2 tens, and 5 ones.

300 has (100) (100) (100) (3 hundreds)

20 has 

(2 tens)

5 has   

(5 ones)

Directions A farmer is feeding his animals. Count to see how many animals are in each group.

1. Horses

____

2. Pigs

____

3. Sheep

____

4. Mice

____

Find the Pattern

Directions Write the missing numbers.

1. 7, ________, 9

2. 18, ________, 20

3. 59, ________, 61

4. 113, ________, 115

5. 12, ________, 14

6. 125, ________, 127

7. 0, ________, 2

8. ________, 43, 44

9. 88, 89, ________

10. 167, ________, 169

Directions Draw the shape that comes next.

11. ● ▲ ● ▲ ● ▲ ________

12. ♥ ■ ■ ♥ ■ ■ ________

13. ▬ ▬ ● ▬ ▬ ● ▬ ________

Animal Riddles

Directions Look at the clues. Think about a word that rhymes with the darker word. Solve each riddle.

EXAMPLE What does a **hen** write with?

A h e n p e n

1. What does a **goat** ride in on the lake?

A g o a t ___ ___ ___ ___

2. What kind of dance does a **pig** do?

A p i g ___ ___ ___

3. What does a **cat** wear on its head?

A c a t ___ ___ ___

4. What do you call a **funny** rabbit?

A f u n n y ___ ___ ___ ___ ___

On My Farm

Directions **If you were a farmer, what animals would you have on your farm? Draw a picture of your farm.**

CHAPTER 4

Animal Homes

Animals live in all kinds of homes.
Birds make nests.
They use twigs from trees.
Snails take their homes with them.
They live in shells.
Rabbits dig holes.
They live under the ground.
Frogs have wet homes.
They live in the water.

Animal Homes

Directions **Use what you have read. Answer the questions.**

1. Where does a rabbit live?

2. How is the frog's home different from the rabbit's home?

3. What is a **twig**?

4. What other animal carries its house on its back like the snail?

5. Where would you most likely find a bird's nest?

ABC Order

REMEMBER

The **alphabet** shows the letters in ABC order. To put words in ABC order, look at the first letter. Which would come first in the alphabet?

A B C D E F G H I J K L M N O P Q R S T U V W X Y Z

EXAMPLE

Ball
Cat
Dog

Animal Line Up

Directions Look at each group of words. Write them in ABC order.

1. nest, bird, twig ____________________

2. wet, frog, home ____________________

In the Woods

Directions Draw lines to make the words from the box below. Write the words in the box shapes.

animal	bear	mat	nut	tree

1. tr **ear**

2. ani **ee**

3. b **mal**

4. m **ut**

5. n **at**

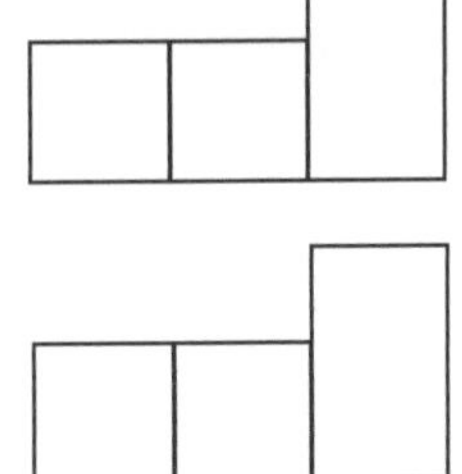

Rhyming Words

Directions Read each word. Find the rhyming words in the box above. Write them on the lines.

6. cut ____________________ ____________________

7. bee ____________________ ____________________

8. hair ____________________ ____________________

9. fat ____________________ ____________________

Counting Coins

REMEMBER

Special signs show the total number of cents. The **¢** sign stands for **cents**. Different coins stand for different amounts of money.

penny 1¢	nickel 5¢	dime 10¢	quarter 25¢

EXAMPLE

   15 cents or 15¢

dime nickel

Directions Circle the coins to show the amount.

1. 8¢

nickel	penny	penny	penny	penny	penny

2. 10¢

nickel	nickel	penny	penny	penny	penny

3. 36¢

quarter	dime	nickel	penny	penny	penny

How Much?

Directions Lee has two groups of coins. Use the coins to answer the questions.

Group A

2 quarters, 3 pennies

Group B

2 dimes, 2 nickels, 3 pennies

1. How many coins are in Group A? ________

2. How many coins are in Group B? ________

3. How many cents do the coins in Group A show? ________

4. How many cents do the coins in Group B show? ________

5. Lee puts the coins from Group A and Group B together. Then he makes two equal groups. Draw a circle around each group of coins.

Tree Codes

Directions **Look at the codes. Each number stands for a letter. Write the letters on the lines to solve the codes.**

A	B	C	D	E	F	G	H	I	J
1	2	3	4	5	6	7	8	9	10
K	L	M	N	O	P	Q	R	S	T
11	12	13	14	15	16	17	18	19	20
U	V	W	X	Y	Z				
21	22	23	24	25	26				

1. How is a tree like an elephant?

__ __ __ __   __ __ __ __
20 8 5 25   2 15 20 8

__ __ __ __   __ __ __ __ __ __.
8 1 22 5   20 18 21 14 11 19

2. What kind of sign does a bird put over its nest?

__ __ __ __   __ __ __ __ __
8 15 13 5   20 23 5 5 20

__ __ __ __
8 15 13 5

3. What did the squirrel think about his new nest?

__ __   __ __ __   __ __ __ __
8 5   23 1 19   14 21 20 19

__ __ __ __ __   __ __!
1 2 15 21 20   9 20

Secret Animal

Think of an animal that lives in the forest. Draw a picture of it. Then write three clues about the animal. Do not name the animal. Read your clues to someone. Have that person guess the animal. Then show them your picture.

1. ______________________________

2. ______________________________

3. ______________________________

CHAPTER 5

What Does a Family Do?

A family may be two people.
It may be more.
A family takes care of you.
They help feed you.
They teach you things.
They play with you.
They read to you.
They make you laugh.
They help you when you feel sad.
A family loves you!

What Does a Family Do?

Directions **Use what you have read. Answer the questions.**

1. What is a family?

2. Circle the sentences from the story that tell about the size of a family.

A family always has four people. There are no more or less.

A family may be two people. It may be more.

3. Name two ways the story says a family takes care of you.

4. What would be another good title for this story?

5. Why is your family special?

Nouns: Naming Words

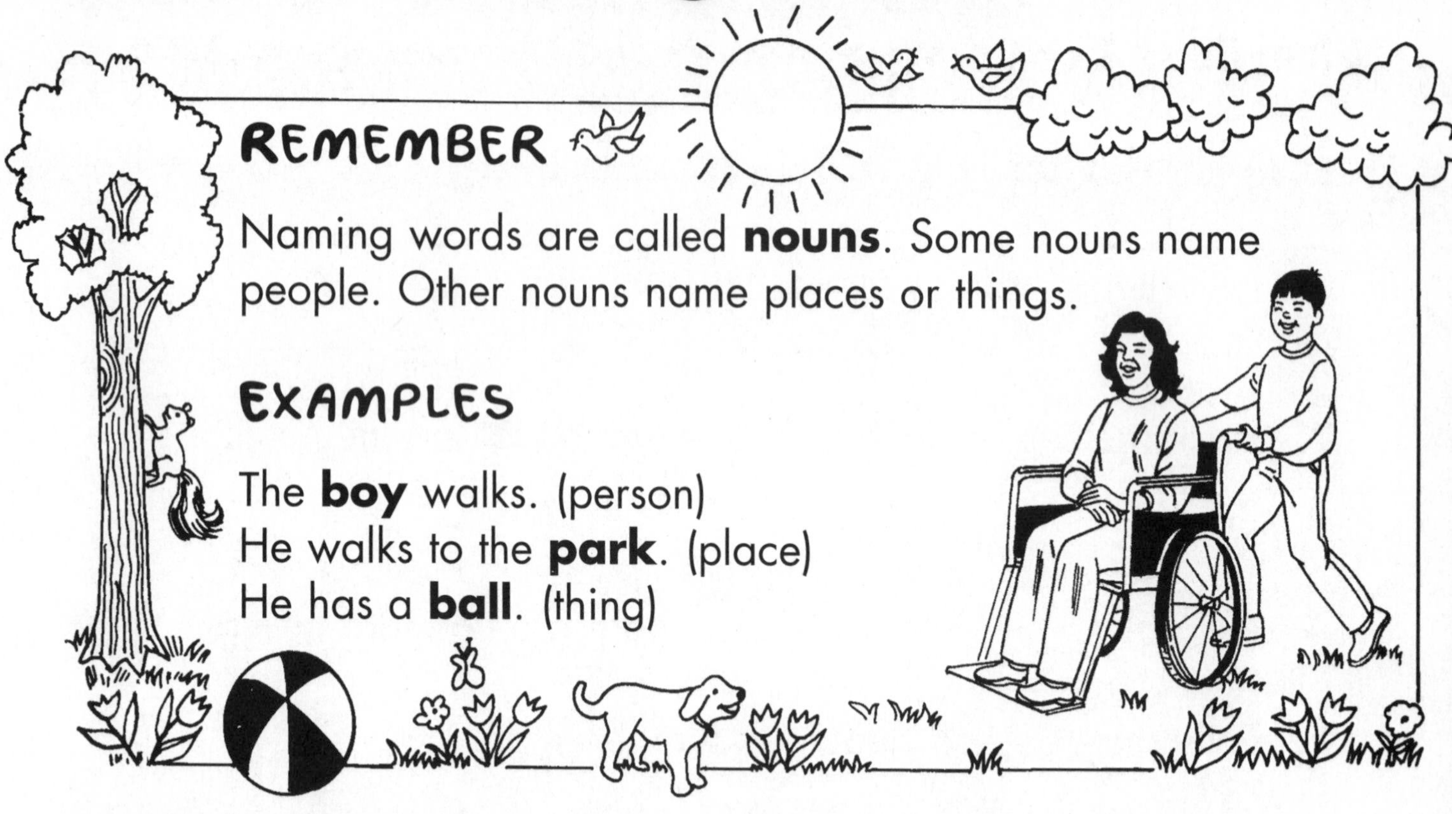

Directions Read each word in the box below. Is the word a person, a place, or a thing? Write it on the correct line.

dad	book	park	girl	dog	school

1. People	**2.** Places	**3.** Things
______	______	______
______	______	______
______	______	______
______	______	______

Picture Clues

Directions **Look at each picture. Find a word in the box below that names each picture. Write it on the line.**

brother	family	Mr.	Mrs.	sister

1. 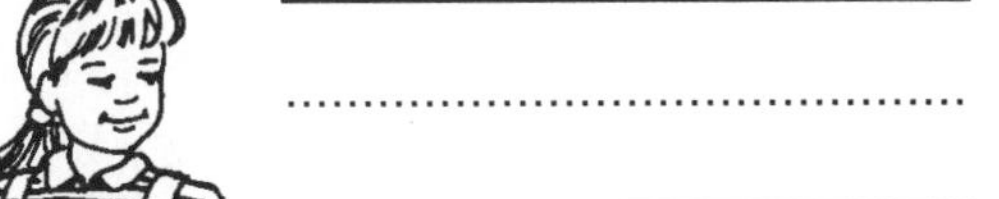 ____________________

2. 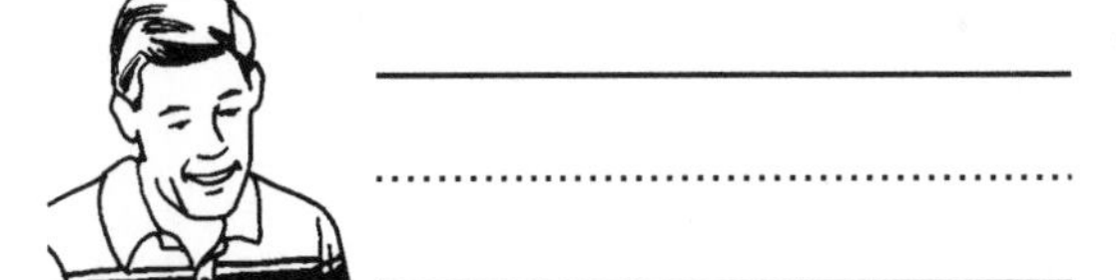 ____________________

3. 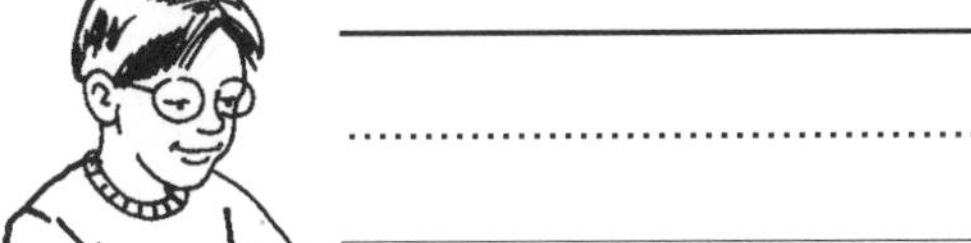 ____________________

4.  ____________________

5.  ____________________

The Right Word

Directions **Read each sentence. Choose words from the box above to finish the sentences. Write them on the lines.**

6. There are five of us in my ____________________.

7. My ____________________ is the tallest girl I know.

8. That boy looks like my ____________________.

Greater or Less Than

REMEMBER

You can compare numbers. Use the signs > and <. The > sign means **is greater than**. The < sign means **is less than**.

EXAMPLES

$5 > 1$ $\quad\quad$ $4 < 5$

5 is greater than 1 $\quad$ **4 is less than 5**

More or Less

Directions **Write < or > in each □.**

1. 1 □ 4 $\quad$ **2.** 12 □ 17

3. 50 □ 7 $\quad$ **4.** 23 □ 20

5. 58 □ 48 $\quad$ **6.** 62 □ 78

7. 85 □ 124 $\quad$ **8.** 201 □ 200

Card Tricks

Directions Paulo has three number cards. Use the cards to complete the page.

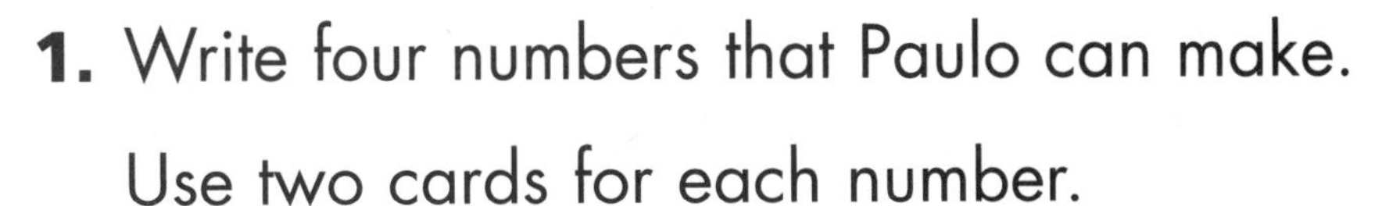

1. Write four numbers that Paulo can make.

   Use two cards for each number.

2. What is the largest number you can make? ________

3. What is the smallest number you can make? ________

4. Write the numbers in order from the smallest to the largest.

   ________, ________, ________

5. What 2 numbers can Paulo make with a two in the ones place?

   ________ and ________

Family Puzzle

Directions **Use words from the box. Complete the puzzle.**

baby	family	sister
brother	mother	father

ACROSS

2. The opposite of sister

3. A man who has a child

4. The opposite of brother

5. A very little child

DOWN

1. A woman who has children

3. A group of people who live together

A Special Person

Directions Is there someone special in your family that you like to be with? What do you do together? Write sentences telling about this person and what you like to do together.

CHAPTER 6

A Vet for Your Pet

Do you have a pet at home?
Maybe it's a dog or cat.
Has your pet ever been sick?
If so, you may have gone to an animal doctor.
An animal doctor is called a vet.
They help animals that are sick or hurt.
Vets help keep animals well, too.
Pets should see their doctor one time each year, just like you!

A Vet for Your Pet

Directions **Use what you have read. Answer the questions.**

1. What do you call an animal doctor?

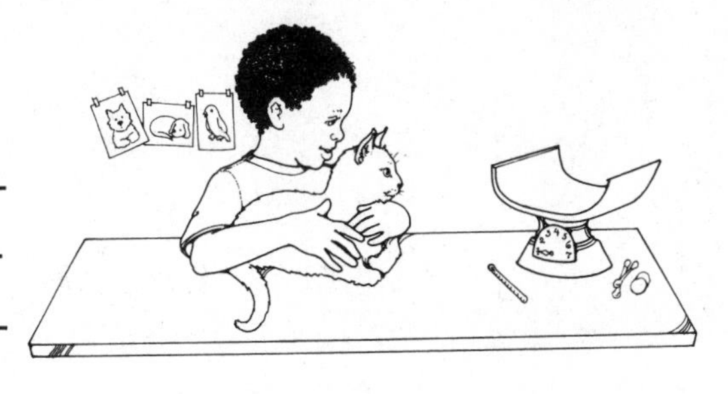

2. How is a vet like a doctor you might visit?

3. How often should you take your pet to the vet?

4. Could this story really happen? Tell how you know.

5. Kahil's cat is not eating. The cat sleeps a lot. What do you think Kahil will do?

Special Nouns

REMEMBER

A **noun** names a person, place, or thing. Some nouns name special people, places, or things. These names begin with capital letters.

EXAMPLES

	Nouns	**Special Nouns**
Person	girl	Maria
Place	street	Brown Road
Thing	day	Friday

Directions Read each sentence. Circle the correct noun.

1. My family lives on (rose road, Rose Road).
2. I have a (dog, Dog).
3. His name is (alex, Alex).
4. We go to the park every (Saturday, saturday) to play.

Pet Words

Directions Look at each picture. Find the words in the box below that name each picture. Write them on the lines.

bird	cat	dog	fish	bunny

My Pet

Directions Read each sentence. Choose words from the box above that best finish the sentences. Write them on the lines.

5. I feed my pet ________________ carrots.

6. I throw a stick to my ________________ Spot.

7. I have a ________________ that swims in a tank.

8. My pet ________________ flies around the house.

Take a Guess

REMEMBER

Some numbers you know. You know that there are five fingers on your hand. Other numbers you must guess. You must guess how old someone is.

EXAMPLE

Is this girl 3 years old? Is this girl 30 years old?

A good guess is that the girl is 3 years old.

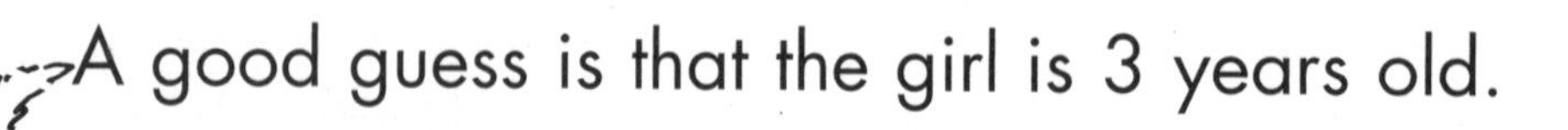

Directions **Circle your guess.**

1. How old? 3 years 7 years 30 years

2. How long? 1 inch 12 inches 1 foot

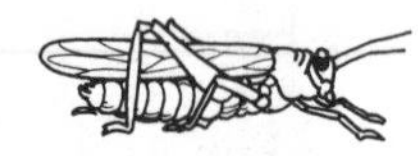

3. How many? 2 beans 20 beans 200 beans

About How Long?

Directions **About how many paper clips long is each item? Circle your answer.**

1.

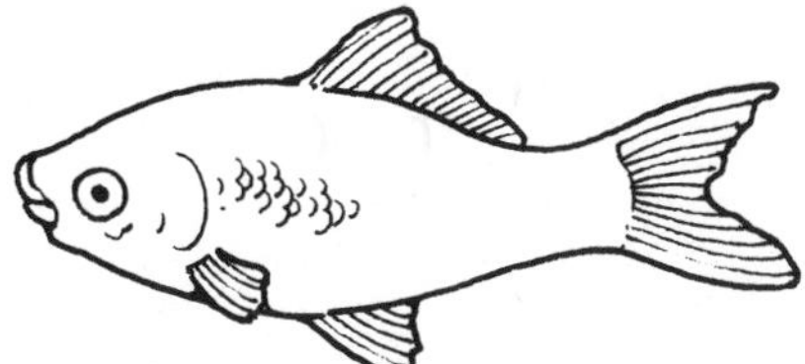

a. 1

b. 2

c. 3

2.

a. 1

b. 2

c. 4

3.

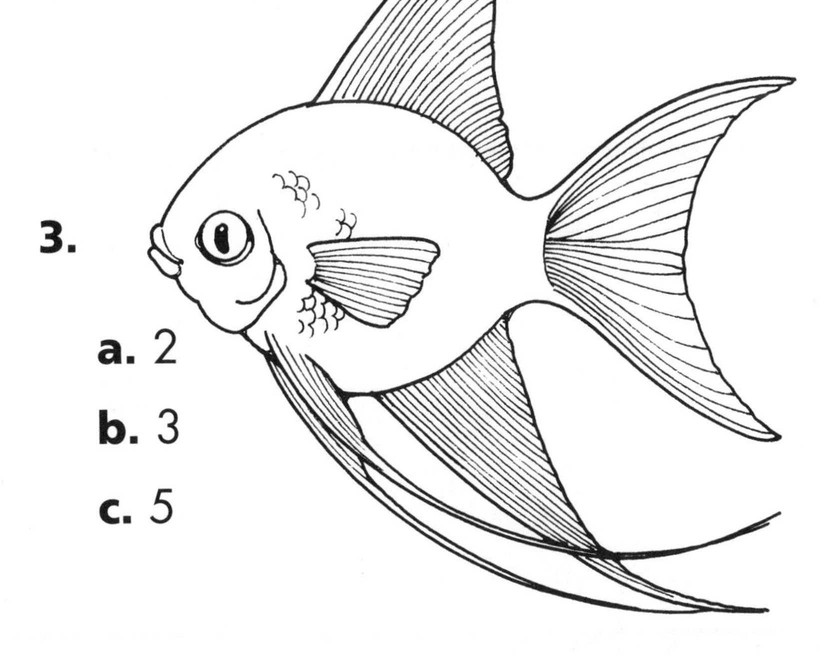

a. 2

b. 3

c. 5

4. About how many paper clips long is the ruler?

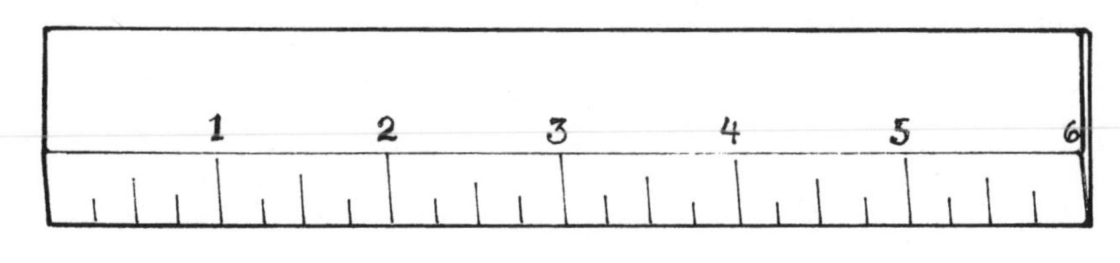

____________ paper clips

Pet Puzzles

Directions **Read each clue. Choose words from the box below to solve each question. Write them on the lines.**

gerbil	kitten	puppy	bird

1. A young dog

___ ___ ___ ___ ___
1 2

2. A pet that can fly

___ ___ ___ ___
3 (under the 3rd line)

3. A pet that looks like a mouse

___ ___ ___ ___ ___ ___
4 (under the 6th line)

4. A young cat

___ ___ ___ ___ ___ ___
5 (under the 5th line)

Directions **Answer the riddle. Match the numbers with the letters. Write the letters on the lines.**

What color do cats like best?

___ ___ ___ ___ - ___ ___ ___
1 2 3 3 1 4 5

Pet Care

Draw a picture of your pet. If you do not have a pet, draw a picture of one that you would like. Think about how to take care of your pet. Write sentences telling how you take care of your pet.

CHAPTER 7

What Is It Like Outside?

What is it like out?
Is it hot?
Some days the sun beats down.
Other days, rain falls.
If it is very cold, it may snow.
Wind can blow.
What will it be like today?
Look up at the sky.
Can you tell?

What Is It Like Outside?

Directions Use what you have read. Answer the questions.

1. What happens when the sun is out?

2. How is a day with sun different from a day with snow?

3. What is happening if leaves shake on trees?

4. If you see dark clouds in the sky, what might happen next?

5. What is it like outside today?

Verbs

REMEMBER

A **verb** can tell what people or things do.

EXAMPLES

The wind **blows**. The sun **shines**.

Rainy Day Verbs

Directions Circle the verb in each sentence.

1. Juan walks home from school.

2. Dark clouds hide the sun.

3. Juan looks at the clouds.

4. Rain falls from the clouds.

5. Juan opens an umbrella.

6. He likes it when it rains.

Weather Words

Directions **Read each word. Find the rhyming words in the box below. Write them on the lines.**

day	rain	sun	wind	cloud

1. run ______________________

2. cane ______________________

3. may ______________________

4. loud ______________________

Which Word?

Directions **Read each sentence. Choose words from the box above that best complete the sentences. Write them on the lines.**

5. A ______________________ covered the sun.

6. The leaves shake when the ______________________ blows.

7. It began to snow the next ______________________.

Measuring Temperature

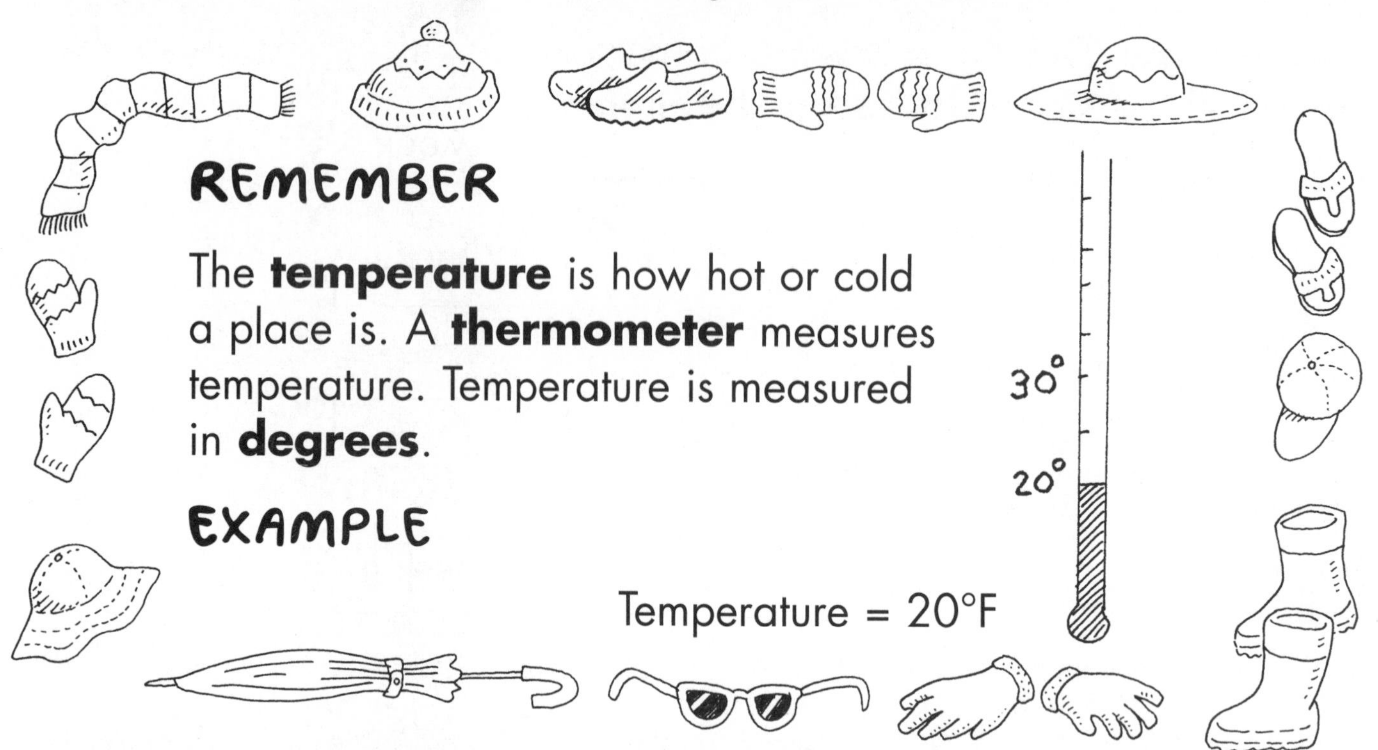

REMEMBER

The **temperature** is how hot or cold a place is. A **thermometer** measures temperature. Temperature is measured in **degrees**.

EXAMPLE

Temperature = 20°F

Directions **Write the temperature shown on each thermometer. Then answer the questions.**

1.

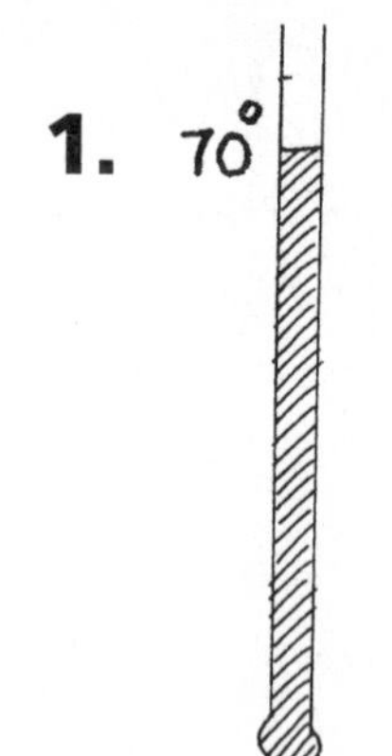

______ °F

2.

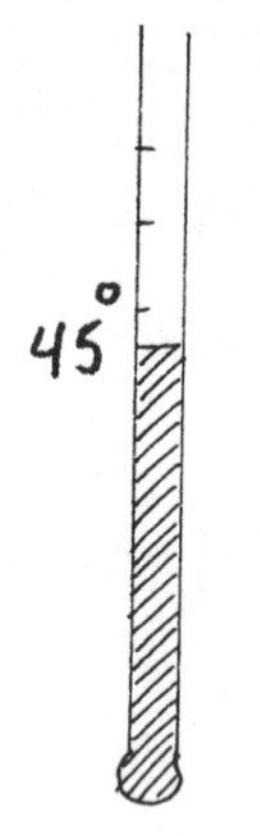

______ °F

3.

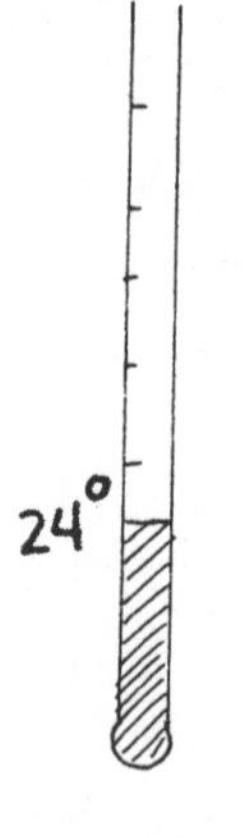

______ °F

4. What is the lowest temperature? ______ °F

5. What is the highest temperature? ______ °F

6. Which temperature is closest to 50°? ______ °F

Sunflower Graph

Directions **The growth of a sunflower is shown on the graph. Use the graph to answer the questions.**

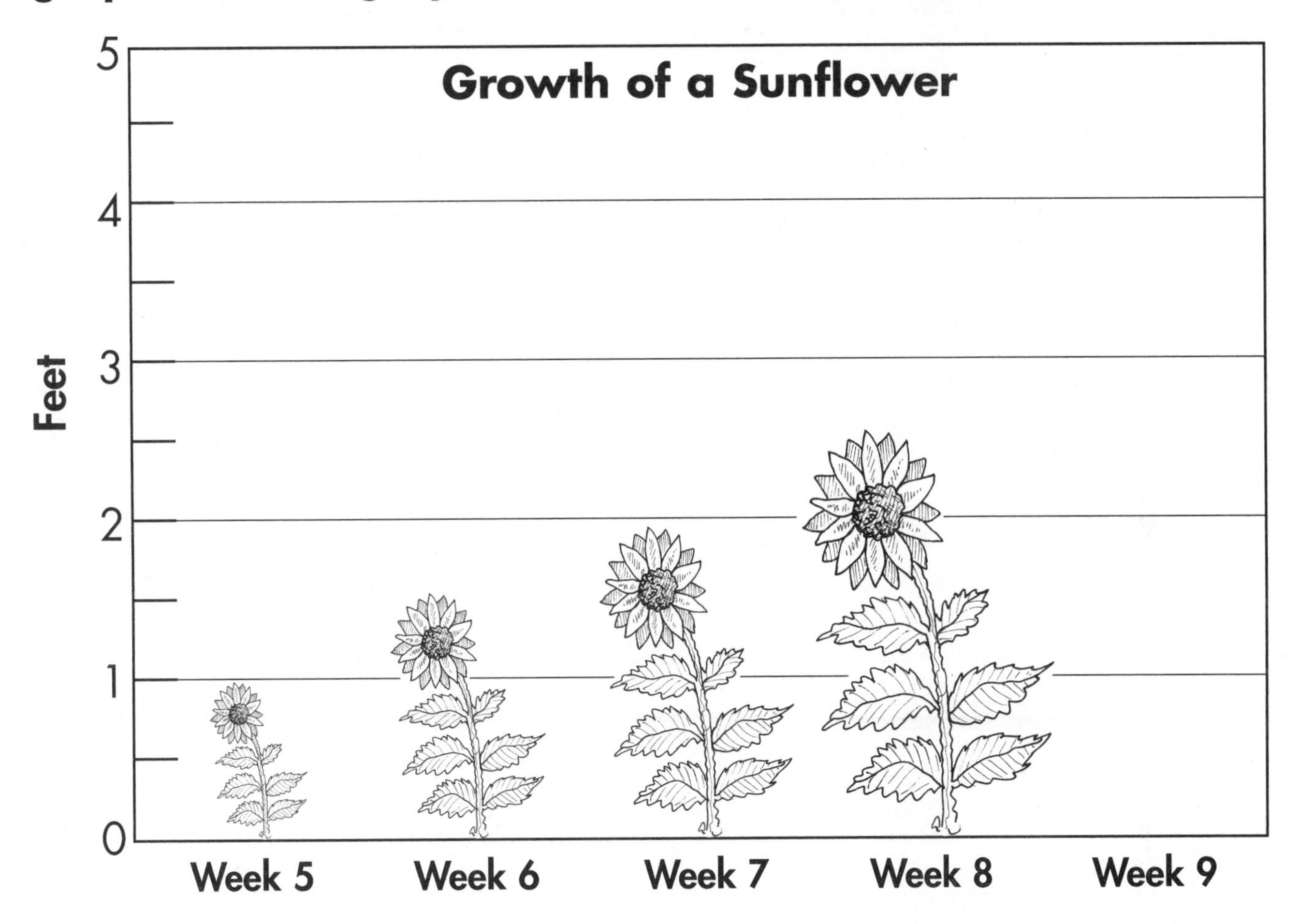

1. How tall was the [sunflower] at week 5? __________ foot

2. How tall was the [sunflower] at week 7? __________ feet

3. How tall do you think the [sunflower] will be at week 9?

__________ feet

4. Draw the [sunflower] on the graph to show its height at week 9.

Weather Puzzle

Directions **Use words from the box. Complete the puzzle.**

hot	sun	cold	snow	wind	cloud

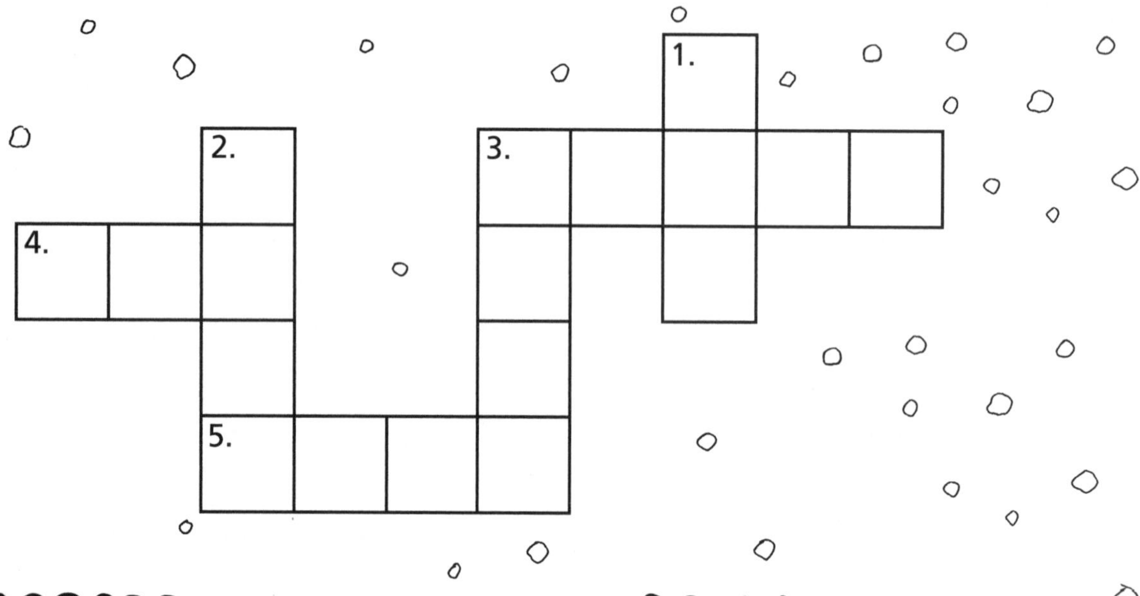

ACROSS

3. Something fluffy that floats in the sky

4. This shines in the sky

5. Something that blows the leaves on the trees

DOWN

1. Very warm

2. Something white that falls from the sky in the winter

3. Not hot

Weather Writing

Look at the picture below. What is happening? Write a story to go with the picture. Give your story a title.

CHAPTER 8

When You Are Together

When you are together with a friend, you share toys.
When you are together, games are more fun.
When you are together, a song might be silly.
When you are together, books seem to be better.
Who are you with?
You're with a friend!

When You Are Together

Directions Use what you have read. Answer the questions.

1. What is this story mainly about?

2. What does **share** mean?

3. Why do you think games are more fun with a friend?

4. What does the story say happens to a song when friends are together?

5. What friend do you like to be with? Why?

Describing Words

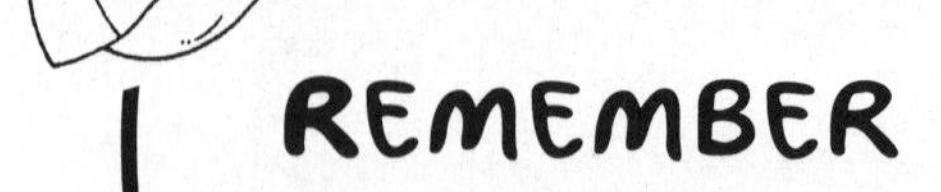

REMEMBER

Describing words tell how something looks, tastes, or smells. They also tell how something sounds or feels.

EXAMPLES

The toy is **soft**.

Rita has a **red** dress.

Playing Catch

Directions Circle the describing word in each sentence.

1. My friend is nice.

2. He has a brown dog.

3. His dog is little.

4. The dog likes to chase a blue ball.

5. The dog is fast.

Friendly Words

Directions Read each word. Find rhyming words from the box below. Write them on the lines.

ask	came	get	read	talk	friend

1. need ______________________ **2.** game ______________________

3. wet ______________________ **4.** walk ______________________

5. mask ______________________ **6.** send ______________________

Friendly Call

Directions Read each sentence. Choose words from the box above that best complete the sentences. Write them on the lines.

7. I ______________________ with my friend on the telephone.

8. I ______________________ her if she will come over to play.

9. She has a new book that we can ______________________.

10. She is my best ______________________.

Shapes and Figures

REMEMBER

A **circle** is a round figure. **Squares** and **rectangles** are figures with four sides. A **triangle** is a figure with three sides. A **dot** can be a figure, too.

EXAMPLES

circle

square

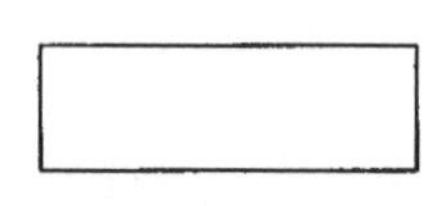

rectangle

triangle

• **dot**

Directions How many figures are used to make each shape? The first one is done for you.

1. 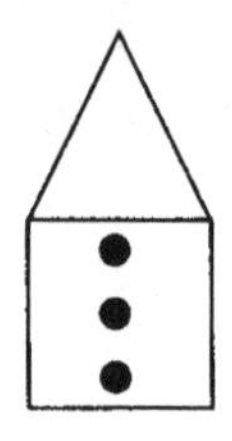

5 ______

2. 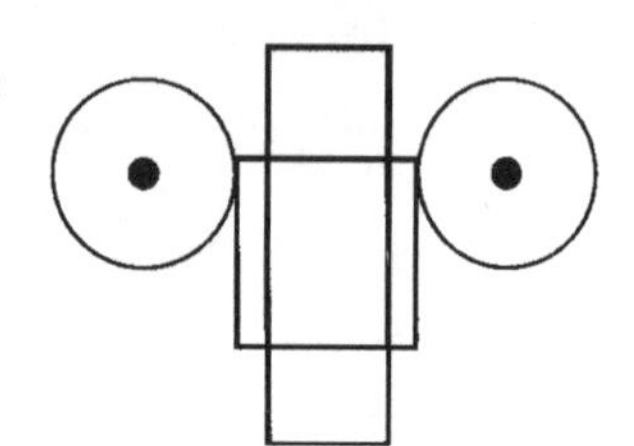

______

3. 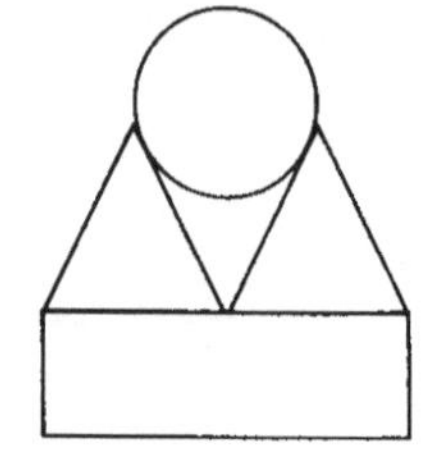

______

4. 

______

5. 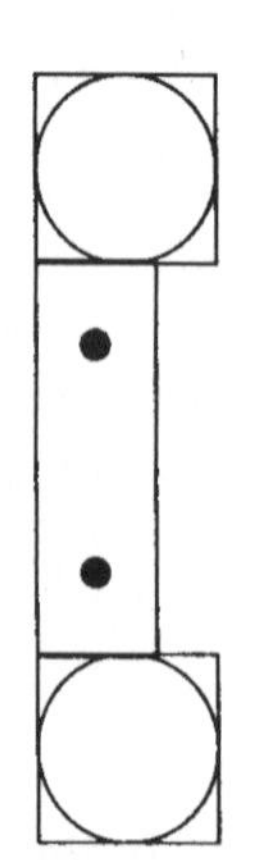

______

6. 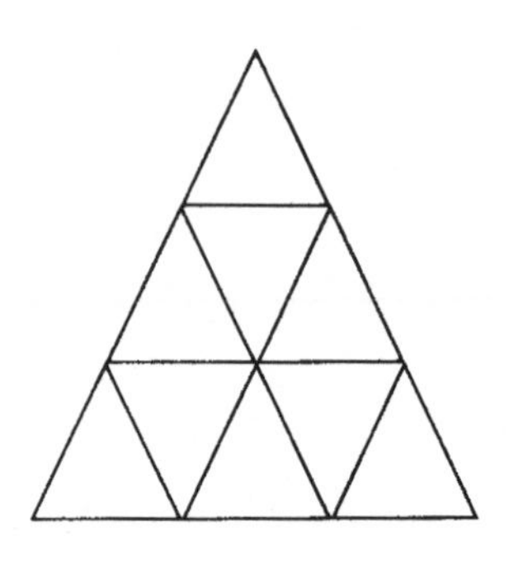

______

Connect the Dots

Directions Follow the directions to answer each question.

1. Connect four dots to make a rectangle.

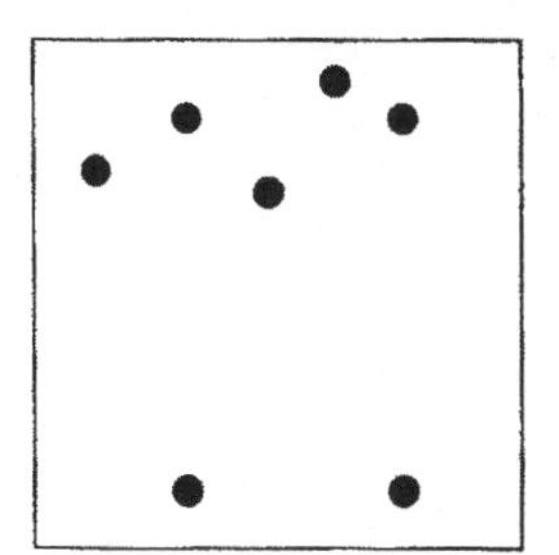

How many dots are outside the rectangle? ____________

2. Connect three dots to make a triangle.

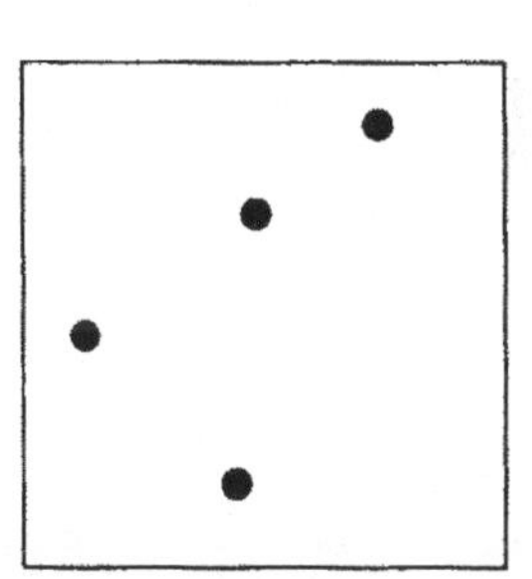

How many different triangles can you make? ____________

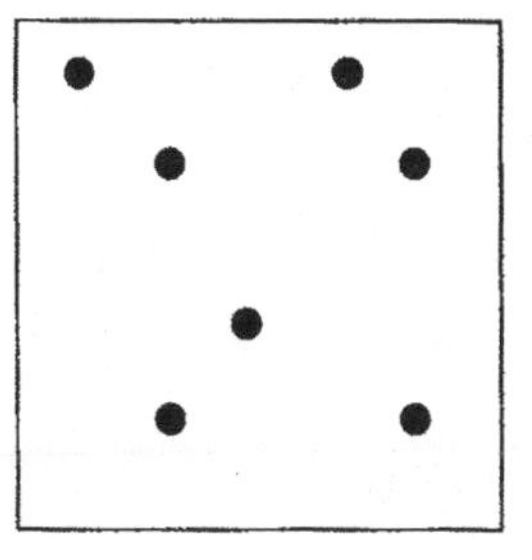

3. Connect four dots to make a square.

Can you make a square using 3 dots? Why or why not?

Find a Friend

Directions Circle the words listed in the box. The words can be across or down.

play	read	sing	share
help	talk	laugh	friend

r	l	p	h	t	a	l	k
n	a	r	e	a	d	b	l
c	u	y	l	s	i	n	g
k	g	s	p	h	c	o	a
p	h	j	g	a	a	g	r
l	y	e	c	r	u	s	s
a	f	r	i	e	n	d	k
y	z	p	m	f	d	n	p

The Perfect Day

Directions **What do you and your friends like to do? Do you like to go to the park? Do you like to play games? Plan your perfect day.**

CHAPTER 9

The LION and the Mouse

One day, Lion was sleeping. Mouse ran up his back. Lion woke up. He gave a big roar. He was angry.

"Who woke me up?" he asked. "I will eat him now!"

"Let me go," said Mouse. "I will help you another day."

Lion laughed. "How could a small mouse like you help me?" he said. But Lion let Mouse go.

Then Lion got caught in a hunter's net. Mouse went to help him. Mouse chewed through the net. Lion was free. Then Lion was glad he let Mouse go.

The Lion and the Mouse

Directions **Use what you have read. Answer the questions.**

1. What is a **roar**?

2. Why was Lion angry when he woke up?

3. Why was Lion glad that he let Mouse go?

4. Write **1**, **2**, and **3**. Tell the order that things happened in the story.

_______ Mouse chewed through the net.

_______ Lion was sleeping.

_______ Lion got caught in a hunter's net.

5. Could this story really happen? Why or why not?

Telling Sentences

REMEMBER

A **telling sentence** tells something.
The sentence begins with a capital letter.
It ends with a period.

EXAMPLES

Lions live in small groups.

Lions hunt for food.

Directions Circle the correct telling sentence.

1. Lions can talk to each other

Lions can talk to each other.

2. they talk by roaring.

They talk by roaring.

3. A loud roar tells other lions to stay away.

a loud roar tells other lions to stay away.

4. That lion wants to be left alone.

That lion wants to be left alone

5. the other lions listen to his roar

The other lions listen to his roar.

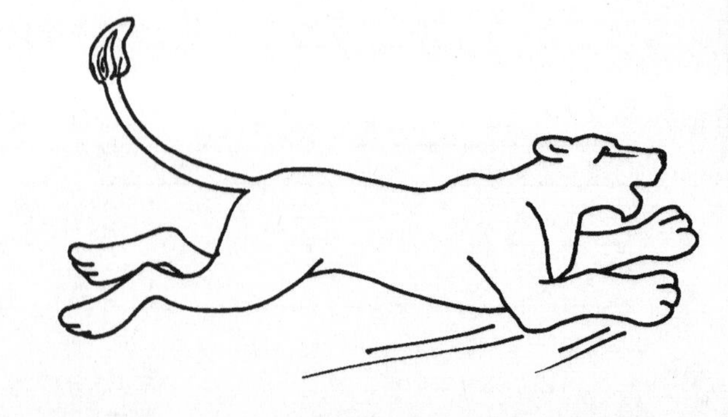

Find the Opposite

Directions Read each word. Find words in the box below that are opposite in meaning. Write them on the lines.

fast	high	little	young	very

1. slow ______________ 2. low ______________

3. old ______________ 4. big ______________

All About Lions

Directions Read each sentence. Choose words from the box above that best complete the sentences. Write them on the lines.

5. Lions are big, but cubs are ______________.

6. Lions can run ______________ for a short time.

7. Lions are ______________ fun to watch.

8. A ______________ lion is called a cub.

Adding Whole Numbers

REMEMBER

When you **add**, you put groups of things together. To show addition, you write **+**. This sign is called a **plus sign** or an **addition sign**. The answer is called the **sum** or **total**.

EXAMPLE

[5 lions] [4 lions] [9 lions]

5 + 4 = 9

Adding It Up

Directions How many shapes are there in all? Write each sum using numbers.

1. ◆ ◆ ◆ + ◆ ◆ = ?

   ______ + _____ = ___

2. ● ● ● ● + ● ● ● ● = ?

   ________ + _________ = ______

3. ■ ■ ■ ■ + ■ ■ ■ = ?

   ________ + ______ = ___

4. ▼ ▼ ▼ ▼ ▼ ▼ + ▼ = ?

   ____________ + ___ = ______

Map Math

Directions **This is a map of where the lions can go. Use the map to answer the questions.**

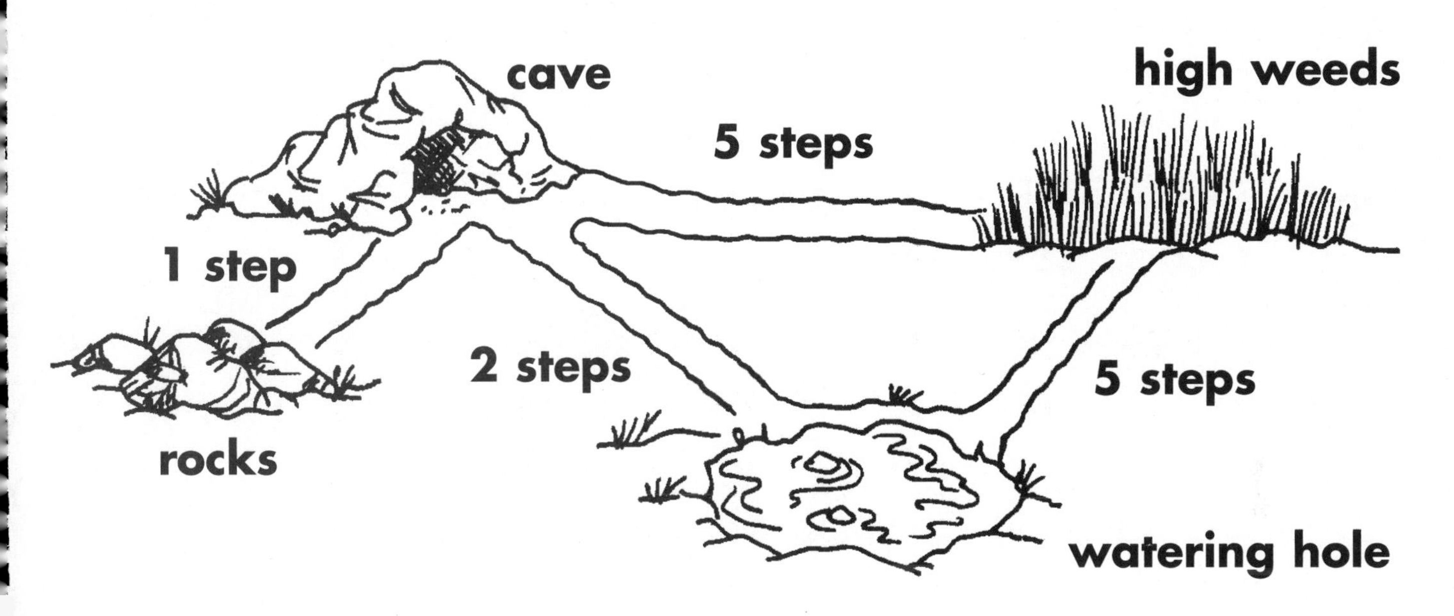

1. What is the least amount of steps from the rocks to the high weeds? _______________ steps

2. What is the least amount of steps from the rocks to the watering hole? _______________ steps

3. A lion cub goes from the rocks to the high weeds. Then she goes to the watering hole. How many steps does the lion cub walk? _______________ steps

4. A lion starts at the rocks and goes to the cave, the high weeds, and the watering hole. Then he goes back to the rocks. How many steps does he walk in all? _______________ steps

Lion Codes

Directions **Look at the code. Each letter has a number. Write the letters on the lines to answer the riddles.**

A	B	C	D	E	F	G	H	I	J
1	2	3	4	5	6	7	8	9	10
K	L	M	N	O	P	Q	R	S	T
11	12	13	14	15	16	17	18	19	20
U	V	W	X	Y	Z				
21	22	23	24	25	26				

1. What group do lion cubs join?

___ ___ ___   ___ ___ ___
20 8 5   3 21 2

___ ___ ___ ___ ___ ___
19 3 15 21 20 19

2. When is a fire like a lion?

___ ___ ___ ___   ___ ___   ___ ___ ___ ___
23 8 5 14   9 20   7 5 20 19

___ ___ ___ ___ ___ ___ ___
18 15 1 18 9 14 7

3. Where do the lions live?

___ ___   ___ ___ ___ ___
15 14   13 1 14 5

___ ___ ___ ___ ___ ___
19 20 18 5 5 20

Helping Out

Directions In the story on page 68, the mouse was able to help the lion even though he was small. Think about a time you were able to help someone even though they thought you were too small. Draw a picture to go with your story.

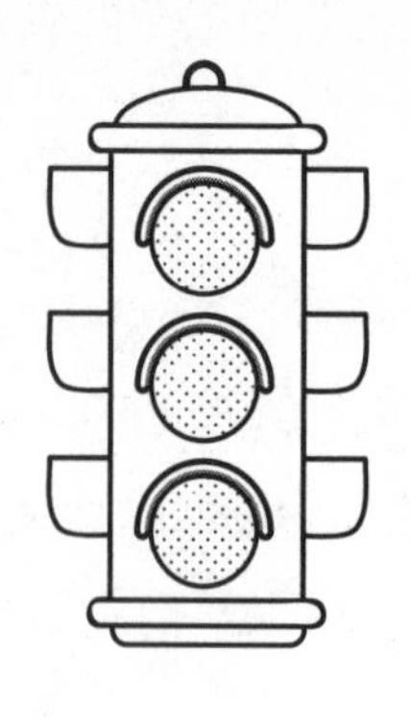

CHAPTER 10

Ride Safely!

Red light—stop! Green light—go!

When people are in cars, they have rules. When you ride a bike, you have the same rules. You have to stop at red lights and at stop signs. You must wait for people to cross the street.

Bike riders have other rules, too. Always wear a helmet. To turn, you must use your hand to show which way you want to go.

Before you go for a ride, make sure your bike is working well. See if both tires have enough air. If you do all these things, you'll be riding right. You'll have a lot of fun too.

Ride Safely!

Directions **Use what you have read. Answer the questions.**

1. What is this story mostly about?

2. How is a person driving a car like a person riding a bike?

3. What is a **helmet**?

4. What would a bike rider do if a person was crossing the street?

5. What is inside a bike tire?

Asking Questions

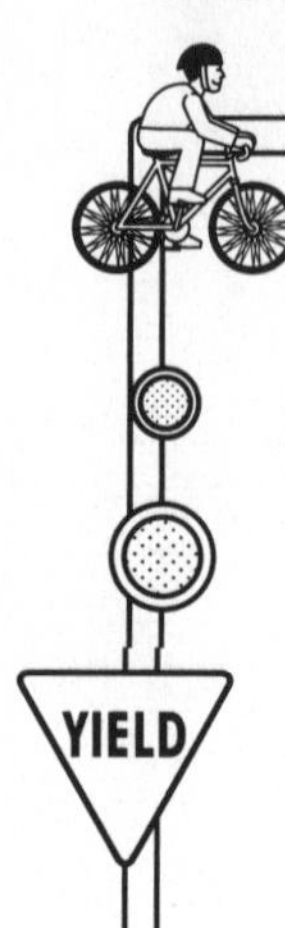

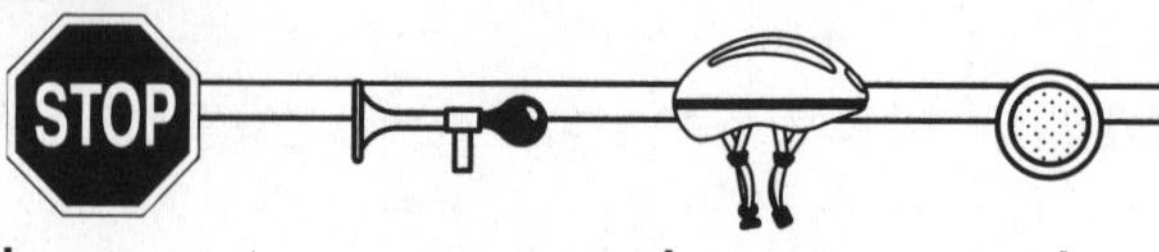

REMEMBER

A **question** asks something. A question begins with a capital letter. It also ends with a question mark.

EXAMPLES

May I play?

What is your favorite game?

Just Asking

Directions **Write each question correctly.**

1. Would you like to play with us

____

2. Which game do you want to play

____

3. is it time for you to go?

____

4. Will you come again

____

Game Words

Directions **Read each word. Find the rhyming words in the box below. Write them on the lines.**

ball	play	ran	ride	toy	game

1. day ____________

2. fall ____________

3. can ____________

4. slide ____________

5. boy ____________

6. name ____________

Just Playing

Directions **Read each sentence. Choose a word from the box above to finish the sentence. Write it on the line.**

7. Would you like to play with my little ____________ car?

8. I like to ____________ my bike.

9. I can hit a ____________ with a bat.

10. It is fun to ____________ with toys.

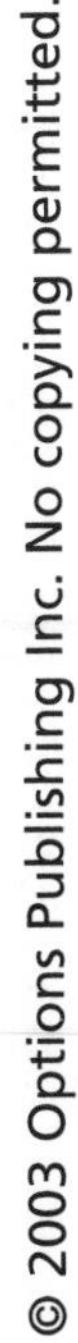

Subtract Whole Numbers

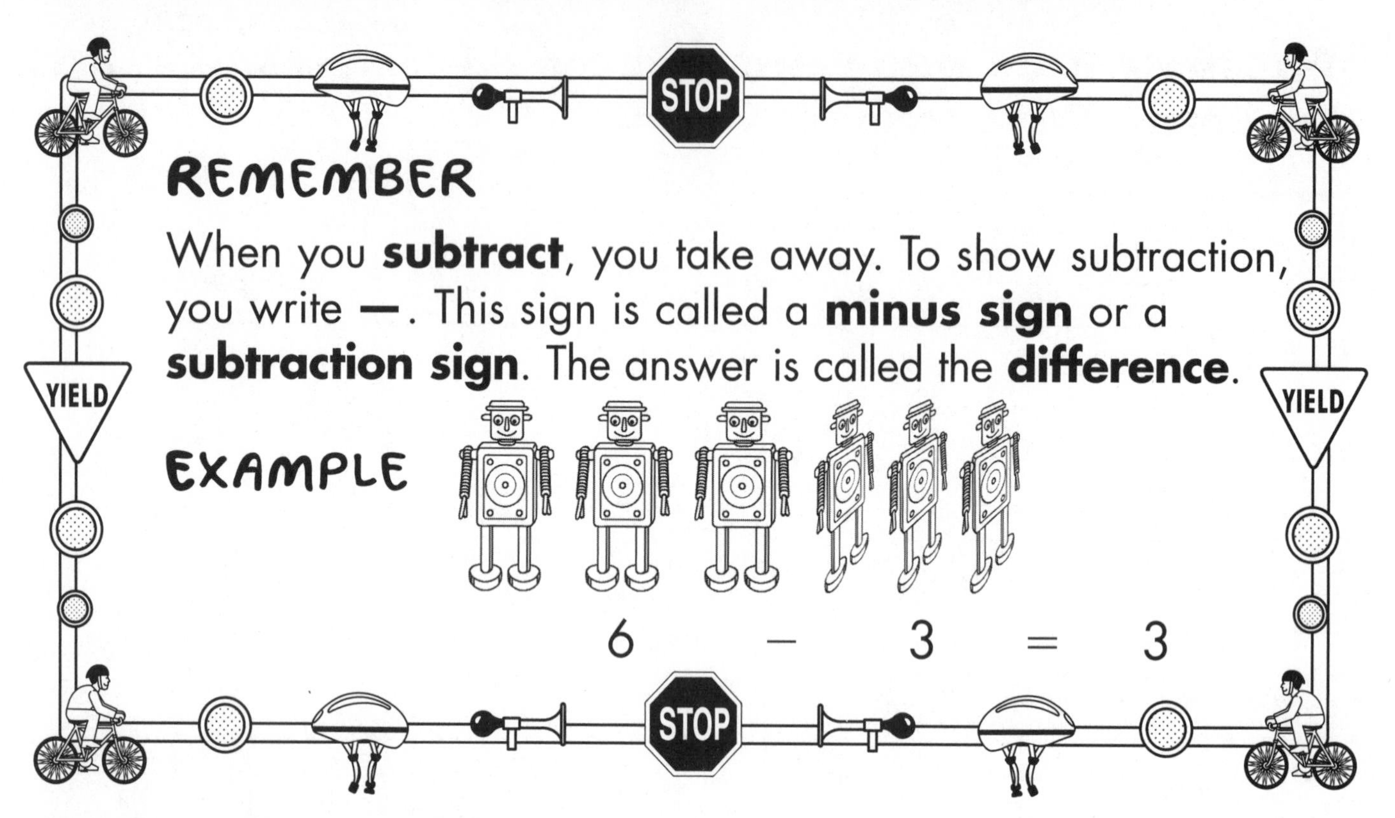

REMEMBER

When you **subtract**, you take away. To show subtraction, you write **—**. This sign is called a **minus sign** or a **subtraction sign**. The answer is called the **difference**.

EXAMPLE

$6 - 3 = 3$

Take It Away

Directions **Some of the toys below are going away. How many are left? Write each difference using numbers.**

1. _____ − _____ = _____

2. _____ − _____ = _____

3. _____ − _____ = _____

4. _____ − _____ = _____

A New Toy

Directions Suppose a toy company asks you to make a new toy. You can use the items below. What will you make? What will your toy do? Draw a picture of your toy. Then write sentences telling about your toy.

  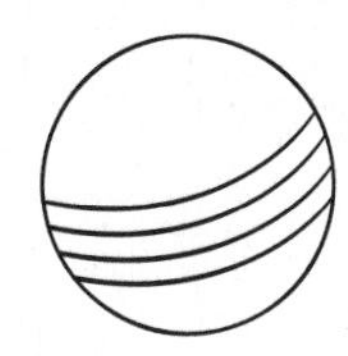  

READING CHECK-UP

What's Happening?

Directions Use the pictures below to answer the questions. Fill in the bubbles next to the correct answers.

1. Jenny won the race.
She came in ______________.

Ⓐ first
Ⓑ last
Ⓒ about
Ⓓ next

2. Tim ______________ out the window.

Ⓐ with
Ⓑ looks
Ⓒ washes
Ⓓ up

3. I drink ______________ when I'm thirsty.

Ⓐ food
Ⓑ under
Ⓒ wind
Ⓓ water

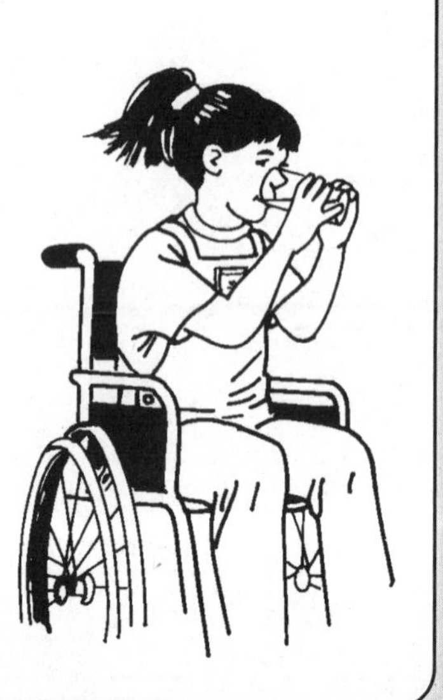

4. The boy sits ______________ the chair.

Ⓐ below
Ⓑ under
Ⓒ near
Ⓓ in

READING CHECK-UP

Swimming

Do you like to swim?
It is fun to move in the water.
Are you fast?
Then you might like to be in a swim race!
Find another swimmer.
Ask someone to say "Go!"
Dive into the water.
Then kick your feet.
Move your arms.
The first one to the end of the pool wins the race.
Good luck!

Directions Fill in the bubble next to the correct answer.

5. What is another good title for this story?

- (A) Run Fast
- (B) Things to Do in the Summer
- (C) A Swim Race
- (D) The Diving Lesson

6. What do you do after you dive into the water?

- (A) Start a swim race.
- (B) Say "Go!"
- (C) Find another swimmer.
- (D) Kick your feet.

READING CHECK-UP

A City Morning

Beep, beep! Ding, ding! Step, step, step!
It's a busy morning in the city.
Many people rush to work.
The streets are full of cars, trucks, and buses.
They take people from place to place.
Stores begin to open.
Some people go in to buy things.
Some people go to the park.
Children play on the grass.
Others walk or jog on the path.
There are many things to do in a city.

Directions Fill in the bubble next to the correct answer.

7. Which of these might go **step, step, step**?

Ⓐ cars

Ⓑ grass

Ⓒ work

Ⓓ people

8. Where would you buy things?

Ⓐ a park

Ⓑ a store

Ⓒ a bus

Ⓓ a path

READING CHECK-UP

Smile, Smile, Smile!

What do you see when you smile? You see your teeth! When you were a baby, you grew teeth. As you got bigger, those baby teeth fell out. New ones grew in. Your teeth help you bite and chew food.

You need to take good care of your teeth. You should brush them two times a day. Also, you should eat foods that are good for you. If you take good care of your teeth, you will have them for a long time.

So now you know about teeth. Give a big smile!

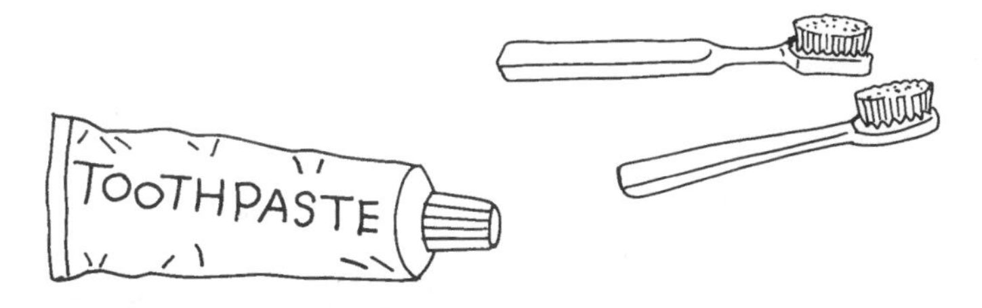

Directions Fill in the bubble next to the correct answer.

9. What happened when your baby teeth fell out?

- Ⓐ New ones grew in.
- Ⓑ You could smile.
- Ⓒ More baby teeth grew.
- Ⓓ No teeth grew in.

10. Which food might be good for your teeth?

- Ⓐ candy
- Ⓑ apples
- Ⓒ soda
- Ⓓ gum

Stop! Number correct: ____ out of 10

Directions **Read each question. Fill in the bubble next to the correct answer.**

1. How many  are in the tank ?

Ⓐ 0 Ⓑ 1 Ⓒ 2 Ⓓ 3

2. What number is missing?

1	2		4	5	6

Ⓐ 3 Ⓑ 4 Ⓒ 5 Ⓓ 6

3. How many  in all?

 +     =

Ⓐ 4 Ⓑ 5 Ⓒ 6 Ⓓ 7

Directions **Read each question. Fill in the bubble next to the correct answer.**

4. What time is it?

Ⓐ 12:00 Ⓑ 1:00 Ⓒ 3:00 Ⓓ 4:00

5. How many cents in a nickel?

Nickel

Ⓐ 1¢ Ⓑ 5¢ Ⓒ 10¢ Ⓓ 25¢

6. Which shape is a square?

Ⓑ 
Ⓒ 
Ⓓ 

Directions **Read each question. Fill in the bubble next to the correct answer.**

7. There are four horses. Two walk away. How many are left?

Ⓐ 2 Ⓑ 3 Ⓒ 4 Ⓓ 5

8. Which group has 5  ?

Ⓐ Ⓑ Ⓒ Ⓓ

9. About how many ants long is the pencil?

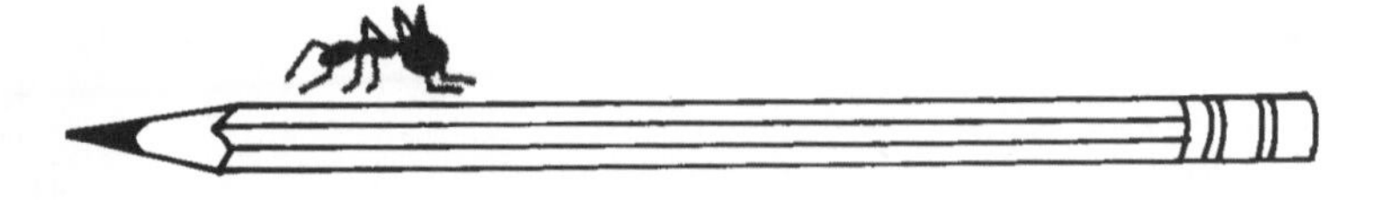

Ⓐ 1 Ⓑ 2 Ⓒ 4 Ⓓ 6

 Stop! **Number correct: ____ out of 9**

Congratulations!

____

(name)

has completed *Summer Counts!*

Good job!

Have a good school year.

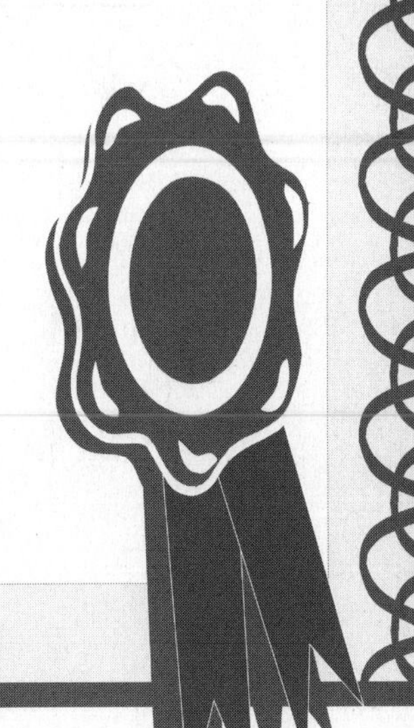

Answer Key

Page 5

1. Possible answer: A fiddle is something that makes music.
2. The cow jumped over the moon.
3. Possible answers: To see such a sport. He was laughing at the cow.
4. No, because cows can't jump over the moon.
5. B

Page 6

1. C
2. I
3. R
4. Y

Page 7

1. love
2. run
3. do
4. see
5. stop
6. do
7. run
8. see
9. stop

Page 8

1. 4
2. 7
3. 3
4. five
5. eight
6. six

Page 9

1. III
2. IIII
3. 𝍸 II
4. 𝍸 I
5. 𝍸
6. 𝍸 III

Page 10

t	s	k	i	p	m	i	l
r	u	n	h	l	r	s	e
r	d	u	d	a	n	c	e
v	l	s	e	y	c	o	j
s	e	h	o	p	a	g	u
i	a	e	c	a	u	w	m
c	p	e	w	a	l	k	p
d	z	p	t	s	h	u	l

Page 13

1. The story is about making pancakes.
2. Answers will vary. Accept reasonable responses.
3. You cook them on a stove.
4. You flip it one time.
5. Answers will vary. Accept reasonable responses.

Page 14

1. p
2. g
3. s
4. b

Page 15

1. eat
2. can
3. box
4. spoon
5. cup
6. box
7. cup
8. eat
9. can

Page 16

1. 11
2. 23
3. 34
4. 45

Page 17

1. 30
2. Sunday
3. Tuesday
4. May
5. Saturday
6. five

Page 18

1. eat
2. can
3. fork
4. dish

Riddle: ears

Page 21

1. The story takes place on a farm.
2. A rooster wakes the farmer.
3. The farmer feeds the pigs first.
4. He goes in to eat.
5. Yes. A farmer feeds his animals. The animals make noises.

Page 22

1. t
2. m
3. o
4. p

Page 23

1. frog
2. sheep
3. cat
4. pig
5. cow
6. cow
7. frog
8. sheep
9. cat

Page 24

1. 111
2. 228
3. 314
4. 432

Page 25

1. 8
2. 19
3. 60
4. 114
5. 13
6. 126
7. 1
8. 42
9. 90
10. 168
11. ●
12. ♥
13. ■

Page 26

1. goat boat
2. pig jig
3. cat hat
4. funny bunny

Page 29

1. A rabbit lives underground.
2. The frog lives in the water and the rabbit lives in the ground.
3. Possible answer: A twig is a small, thin part of a tree.
4. Possible answer: A turtle carries its home on its back like a snail.
5. Possible answer: You would find a bird's nest in a tree.

Page 30

1. bird, nest, twig
2. frog, home, wet

Page 31

1. tree
2. animal
3. bear
4. mat
5. nut
6. nut
7. tree
8. bear
9. mat

Page 32

1. 1 nickel and 3 pennies
2. 2 nickels
3. 1 quarter, 1 dime, 1 penny

Page 33

1. 5
2. 7
3. 53¢
4. 33¢
5. Each group should have 1 quarter, 1 dime, 1 nickel, 3 pennies.

Page 34
1. They both have trunks.
2. Home tweet home
3. He was nuts about it!

Page 37
1. Possible answer: A family is a group of people who take care of each other.
2. A family may be two people. It may be more.
3. Possible answers: A family helps feed you, they teach you things, they play with you, they read to you, they make you laugh, they help you when you feel sad.
4. Answers will vary.
5. Answers will vary. Accept reasonable responses.

Page 38
1. dad, girl
2. park, school
3. book, dog

Page 39
1. sister
2. Mr.
3. brother
4. Mrs.
5. family
6. family
7. sister
8. brother

Page 40
1. <
2. <
3. >
4. >
5. >
6. <
7. <
8. >

Page 41
1. Possible answers: 27, 25, 75, 57, 52, 72
2. 75
3. 25
4. 2, 5, 7
5. 72 and 52

Page 42
Across
2. brother
3. father
4. sister
5. baby

Down
1. mother
3. family

Page 45
1. A vet is an animal doctor.
2. Possible answer: Both help sick people or animals.
3. You should take your pet to the vet once a year.
4. Yes, because a vet is a real person who takes care of animals
5. Possible answer: Kahil will take his cat to a vet.

Page 46
1. Rose Road
2. dog
3. Alex
4. Saturday

Page 47
1. cat
2. dog
3. bird
4. fish
5. bunny
6. dog
7. fish
8. bird

Page 48
1. 7 years
2. 1 inch
3. 200 beans

Page 49
1. c
2. b
3. c
4. 6

Page 50
1. puppy
2. bird
3. gerbil
4. kitten

Riddle: purr-ple

Page 53
1. Possible answer: It gets hot.
2. Possible answer: The sun makes it hot out. The snow makes it cold out.
3. The wind is blowing.
4. Possible answers: It might rain or snow.
5. Answers will vary. Accept reasonable responses.

Page 54
1. walks
2. hide
3. looks
4. falls
5. opens
6. likes

Page 55
1. sun
2. rain
3. day
4. cloud
5. cloud
6. wind
7. day

Page 56
1. 70
2. 45
3. 24
4. 24
5. 70
6. 45

Page 57
1. 1 foot
2. 2 feet
3. Possible answer: 3 feet
4. Drawing should show sunflower at about 3 feet tall.

Page 58
Across
3. cloud
4. sun
5. wind

Down
1. hot
2. snow
4. cold

Page 61
1. The story is about how things are more fun when they are done with a friend.
2. Share means to let another person use something of yours.
3. Answers will vary. Accept reasonable responses.
4. Songs might be silly.
5. Answers will vary. Accept reasonable responses.

Page 62
1. nice
2. brown
3. little
4. blue
5. fast

Page 63
1. read
2. came
3. get
4. talk
5. ask
6. friend
7. talk
8. ask
9. read
10. friend

Page 64
1. 5
2. 6
3. 4
4. 6
5. 7
6. 9

Page 65
1. 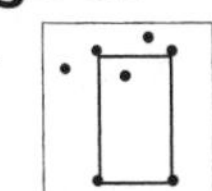

There are 2 dots outside the rectangle.

2. 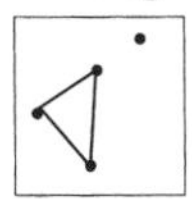

You can make 3 triangles.

3. 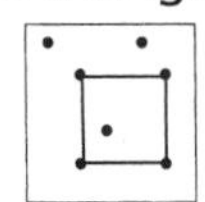

A square must have 4 sides, so it will need four dots.

Page 66
r	l	p	h	t	a	l	k
n	a	r	e	a	d	s	l
c	u	y	l	s	i	n	g
k	g	s	p	h	c	o	a
p	h	j	g	a	a	g	r
l	y	e	c	r	u	s	s
a	f	r	i	e	n	d	k
y	z	p	m	f	d	n	p

Page 69
1. A roar is the loud sound a lion makes.
2. Possible answer: Lion was angry because he did not want to be woken up.
3. Lion was glad because Mouse was able to help him.
4. 3, 1, 2.
5. No, because animals don't talk in real life.

Page 70
1. Lions can talk to each other.
2. They talk by roaring.
3. A loud roar tells other lions to stay away.
4. That lion wants to be left alone.
5. The other lions listen to his roar.

Page 71
1. fast
2. high
3. young
4. little
5. little
6. fast
7. very
8. young

Page 72
1. $3 + 2 = 5$
2. $4 + 4 = 8$
3. $4 + 3 = 7$
4. $6 + 1 = 7$

Page 73
1. 6
2. 3
3. 11
4. 14 steps or 22 steps, if he takes the same route.

Page 74
1. The cub scouts
2. When it gets roaring
3. On Mane Street

Page 77
1. The story is about riding a bike safely.
2. They both must follow rules.
3. A helmet is a hard hat worn on the head to keep safe.
4. Possible answer: The bike rider would stop.
5. Possible answer: Air is inside a tire.

Page 78
1. Would you like to play with us?
2. Which game do you want to play?
3. Is it time for you to go?
4. Will you come again?

Page 79
1. play
2. ball
3. ran
4. ride
5. toy
6. game
7. toy
8. ride
9. ball
10. play

Page 80
1. $3 - 1 = 2$
2. $5 - 2 = 3$
3. $4 - 2 = 2$
4. $7 - 4 = 3$

Page 81
1. 9:00
2. 9:30

3. 8:45

4. 10:00

Page 82
1. toy joy
2. ball wall
3. red sled
4. book look

Page 84
1. A
2. B
3. D
4. D

Page 85
5. C
6. D

Page 86
7. D
8. B

Page 87
9. A
10. B

Page 88
1. C
2. A
3. B

Page 89
4. D
5. B
6. A

Page 90
7. A
8. C
9. D